INTRODUCTION

In this, the fourth 'Vintage Steam Album', we come to those popular Steam ... as is demonstrated when one comes up for sale in a reasonable condition. It is usually snapp... indeed.

Many of the tractors we see today are engines built by Aveling & Porter, or Richard Garrett & Sons Ltd.; both companies built large numbers over the years. Garretts supplied a considerable number to the War Department during the 1914-18 war, many of which were offered for sale after the end of hostilities; a sizeable number of these very engines have survived into preservation.

The other large engine builders are represented by such engines as the Burrell 'Gold Medal Tractor', the Fowler 'Tiger' and the Foster 'Wellington'; also the Tasker 'Little Giant' tractor. The Foden company have several types including the 'D' type and two examples of the 'Sun' tractor. Many of the tractors built by Wallis & Steevens are of the 'Oil-bath' type. In addition some companies are represented by single figures; notable amngst these are the two Robey 'Express' tractors - an unusual and interesting design. The sole Yorkshire tractor was originally used as the tractor unit of an articulated vehicle and operated by an electricity company.

During their working days tractors were to be seen hauling a wide variety of loads ranging from coal, stone, bricks, and even furniture. If the tractor was under five tons in weight they were usually operated by one man, thus providing the operator with considerable savings.

The Sentinel company of Shrewsbury built several designs of tractors, ranging from the powerful timber tractors such as 'Brutus', an eleven ton unit built in 1933, which is fitted with two 12NHP engines working from the boiler, one providing propulsion, while the other is used to drive a winch. Note many of these 'Timber tractors' were built; they were also quite expensive when new. A small number have survived into preservation. 'Brutus' is now part of the famous Bressingham collection.

One tractor unit which was a familiar sight on the rally field during the seventies was named 'The Elephant', an unusual 'Super' Sentinel, works number 5644, built in 1924. For many years this Sentinel was used to shunt railway wagons on Teignmouth Quay, Devon. This Sentinel was sold in 1989 to a Dutch enthusiast.

Other Sentinels are to be seen with matching trailers, one of which has made a Continental tour in recent years attracting much attention during the journey, whilst others were converted from Wagons in the early days of preservation due to taxation difficulties. A Wagon had to carry a full licence, whilst a tractor could be taxed with an 'F' or agricultural licence.

The Mann Patent Steam Cart & Wagon Co. Ltd., of Hunslet, Leeds produced an unusual design known as the 'Mann Steam Cart', being in fact a tractor designed for operation by one man. The firebox door was sited so that the driver could maintain the fire. In fact, the driving position was the only space on these unusual engines. Several options were available including a cart body, although the engines were primarily designed for agricultural use including direct ploughing. Nine tractors built by this company are in preservation.

Other Companies with a small number of tractors surviving are McLarens with three. Marshalls of Gainsborough, Lincolnshire have many engines surviving, very few of which are tractors. Ransomes, Sims & Jefferies the famous Ipswich company have just six survivors in this country. Finally, Ruston Proctor and Ruston & Hornsby have the same number in total.

The showmans tractors have not been included in this title; these will be dealt with separately, as will the much larger road locomotives.

Mann 4NHP tractor no. 1260 was built in 1917 and is thought to have been supplied to a Royston, Hertfordshire farmer where it was used for direct ploughing, threshing, haulage, etc. These tractors were the equivalent of todays modern internal combustion tractors. Photographed at Harewood House, Leeds, rally in 1971.

1. This superb tractor was built by Aveling & Porter in 1920 as work number 9183. In 1973 it attended Expo steam at Alwalton, Peterborough, where this photograph was taken. Note the 'Invicta' badge carried on the belly tank, and the distinctive cast wheels used on Aveling tractors.

2. Aveling & Porter tractors are capable of a fair turn of speed. This 5 ton tractor is number 12152 and was built in 1928. It was photographed at Weeting, an event it attended on several occasions. The compound cylinder block has outside valves, and there is an external exhaust steam pipe into the smokebox. This engine has piston valves as indicated by the round sides to the cylinder block.

3. An earlier Aveling & Porter 4NHP tractor 8809 built in 1917 and named 'Flower', photographed at the 1963 Chatteris rally. This compound tractor weighs six tons.

4. Another photograph of Aveling & Porter 'Flower' number 8809; this time at the 1961 Ickleton rally. During the sixties this engine attended many East Anglian events.

5. 'Emma' quitely raises steam ready for the days activities at the 1967 Market Bosworth rally. The Aveling & Porter is works number 9267, built in 1920. Note the effect on the engine's appearance when fitted with a canopy. This engine is in fact a 'convertible'.

6. Another view of Aveling & Porter tractor 'Emma' showing the other side of the engine. Weighing in at eight tons this engine is heavier than the rather similar Garretts. Photographed at Rempstone in July 1966, this engine is owned by the Hentons of 'Hogs Norton' in Leicestershire whose name will be familiar to those who have read the "Field Marshall" title from this publisher!

7. This Aveling & Porter tractor started its working life for Hereford County Council and ended its working days like so may other engines on agricultural duties. The engine is works number 9228, built in 1921, and is a 5NHP compound slide valve engine weighing 6.5 tons. Slide valve engines had the flat sides to the cylinder block.

8. Aveling & Porter tractor number 11486 was built in 1926 and supplied to an engineering company. Here this engine is seen at the 1967 Pirton rally, on a rather gloomy day. The rather large cylinder block with piston valves was a part common to many of the same manufacturers Rollers.

9. 'Oberon' is a 4NHP Aveling & Porter tractor built in 1927 as works number 11839, photographed here on a damp, dismal day at Castle Howard.

10. Aveling & Porter tractor number 11997 was built in 1928. 'Lucy May' was photographed at the Thetford rally in July 1967. A large number of these popular tractors were built by Avelings over the years.

11. We now move from Thetford rally to the products of Thetford itself. Burrell 'Gold Medal' tractor number 4072 'Tinkerbell' was built at Thetford in 1927. This engine was attending the Pirton Herts rally in 1974 when this photograph was taken.

12. Burrell tractor 'Sunrise' works number 3689 built at Thetford in 1915. This engine is seen while attending the Great Wymondley rally held near Hitchin, Herts., in June 1965.

13. Burrell number 3554 'King George V' was built in 1914, the front tyres being pneumatic giving a rather unusual appearance to this Burrell.

14. 'Furious' makes its way to the ring at the 1966 Great Wymondley rally. This Burrell is works number 3191 built at Thetford in 1910. 'Furious' is one of the earliest Burrell tractors to survive. Most Burrell tractors were double crank compounds, which was the usual layout for most steam tractors.

15. The only Clayton & Shuttleworth tractor known to survive is this one 'Appollo', works number 49008 built at Lincoln in 1924. It was 'works engine' for many years until sold some ten years later.

16. In 1986 Foden tractor number 11444 'Hielan Laddie' made the long journey from its Aberdeen base to attend the Great Dorset Steam Fair'. This superbly restored Foden was built at Sandbach in 1924 and is typical of the tractors from this concern based on their overtype steam wagon design. Indeed some engines now appearing as tractors started life as wagons.

17. This unusual Foden tractor is an example of the companies two cylinder 'Sun' design, built in 1931. The tractor's working parts can clearly be seen in this photograph and these included a totally enclosed motion. One other example of this design is in the country having returned from overseas. Unfortunately this most interesting engine does not attend many rallies.

18. The other side of Foden number 13730 showing the flywheel and steering position. Foden tractors and wagons always attract attention at rallies especially this unusual and superbly restored example of the 'Sun Tractor'.

41. Only three tractors built by McLarens of Leeds have survived, this one, number 1413 'May Queen', was built in 1913 and is the oldest of those surviving, seen here while attending a Weeting Rally.

42. Another picture from the Rally archives, this time of McLaren number 1837 'Bluebell' photographed at Raynham in September 1963. Tractors built by several companies were to be seen at this event.

21. Foden 13222 is a fine example of Foden's D3 three speed slide valve compound tractor. This engine is widely travelled and draws attention wherever it goes. Of note is the Ackerman steering which is fitted to most Fodens

22. The flywheel side of Foden 13222 'Cheshire Maid'. The tractor was built in 1928 and is typical of their 4NHP design.

23. This 4NHP Foden tractor number 13218 'Cestria' was built in 1928; like so many others, this engine worked for part of its life at a sawmill.

24. Foden number 12770 was built at Sandbach in 1927, starting life as a wagon and since being converted into a tractor, as were many other examples. This photograph was taken in May 1990 at Eastnor Castle, Hereford.

25. Foden tractor number 13068 'Perseverance' was built at Sandbach in 1928. For many years it was used for timber haulage in South Wales, during the last war this Foden assisted in clearing bombed buildings in London and ended its working life in the county of Essex.

26. Foden tractor number 13762 'The Dalesman', built in 1931, is the subject of discussion at the 1967 Castle Howard, Yorkshire, rally. This photograph shows clearly the design of the wheels.

27. In the sixties this 4NHP tractor was a familiar sight on the rally fields. The engine is works number 13196 and is named 'Pride of Fulham', having been originally operated by a London coal merchant. Note the solid rubber tyres all round and rear hub caps which read 'Foden Six Tonner'.

28. The Raynham rallies were held near Fakenham in 1962 and 1963 and attracted engines from a wide area, including Foden 13196 which had travelled from Leicestershire to attend the September 1963 event.

29. Foster tractor 13031 was built at Lincoln in 1913 and carries the name 'Mighty Atom'. It was photographed at Rempstone in July 1966. Not many Foster tractors have survived, and these include some which are Showmans tractors, a subject which will be covered in a separate volume in this series.

30. Another rally scene taken over twenty five years ago. This picture shows Fowler Tiger Tractor 15629 built at Leeds in 1920. This photograph was taken at Great Wymondley in June 1965. The engine is still to be seen regularily at rallies in eastern England.

31. This fine example of the Fowler T3 'Tiger' tractor is number 15632 built at Leeds in 1923; a nippy tractor which over the years has travelled to and from a great many events under its own steam.

32. Fowler T3 tractor number 14406 'Pandora' was built in 1917, and was photographed at Stamford in 1966. This engine was very widely travelled during the sixties, travelling on a converted Crossley bus fitted with a novel turntable transporter.

33. Garrett 'Patricia' photographed at Great Wymondley in 1965. This engine is works number 33991, built at Leiston in 1921. 'Patricia' has been a regular on the rally scene for many years.

34. This Garrett tractor is one of the batch built in 1918 for the War Department. In 1922 it was purchased by a Birmingham based showman and converted to a showmans tractor. It was later converted to a haulage tractor. The engine still has reminders of its showland days with brass nameplates and nickel plated items. 'Princess Mary' is works number 33278, a fine example of the Garrett 4NHP tractor.

35. Garrett number 33295 'Princess Royal' photographed at the 1989 Weeting rally. Note the twisted brasswork on this engine which is a reminder of her showland days when she was operated by Henry Thurston in the 1920s. Unfortunately no photograph has come to light of this engine in showland use when it carried the name 'Felix'. Should any reader know of one please contact the author. This engine was another originally built for the War Dept. in 1918.

36. 'Dorothy' is a fine example of the Garrett 4NHP tractor built in 1920 as works number 33782. The engine was photographed at the 1989 Weeting rally. When new this engine was supplied to Flowers & Sons, Reading.

37. This Garrett tractor has a long record of attending rallies with over thirty years completed. Number 33141 was built at Leiston in 1918. Note the Garrett emblem mounted on the chimney.

38. Only nine tractors built by the Mann Patent Steam Cart & Wagon Co. Ltd., remain in preservation, this one is number 1425, built at Leeds in 1920. The design was primarily for direct ploughing, as can be seen in the companion volume in this series on Ploughing Engines, although it had many other applications.

39. Marshall 5NHP tractor number 36258 was built in 1902 and started life as a roller. When this photograph was taken it was still undergoing restoration. A large number of Marshalls are in preservation but they include very few tractors.

40. Marshall tractor number 36258 as it is today; this is a recent photograph taken in May 1990 and shows the amount of work put into restoration, and the superb end result attained. This veteran Marshall left the Gainsborough works in 1901.

19. This example of the Foden 'D' type tractor number 14078 'Mighty Atom' was built in 1932 and was supplied new to a Timber Merchant; in due course it found employment with Coles roundabouts. Extensive work has been carried out on this engine and it is now to be seen in superb condition.

20. For many years this six wheeled tractor was owned by the Limmer & Trinidad Asphalt Co., based in Liverpool and used for tar spraying duties. When photographed the engine carried the name 'Wanderin Tam'. Since then this engine, works number 13008, built in 1928, has changed hands and has moved south of the border, regularly attending events in many parts of the country.

43. Examples of two well known builders 4NHP tractors. In front is McLaren number 1837 'Bluebell' built at Leeds in 1936; the other is Garrett 33380, built in 1918 and named 'Sapphire'. This engine now carries showmans fittings. The picture was taken at the Weeting rally in July 1985.

44. This neat Ransomes, Sims & Jefferies tractor is the oldest surviving tractor built by the Ipswich company. The works number is 23266, built in 1911 and named 'Backus Boy'. The photograph was taken way back in September 1962 while attending the Raynham rally.

45. Robey & Co. Ltd., of Lincoln built nine of these unusual 'Express Tractors', of which only two are thought to have survived. Works number 43388 was built in 1929. The engine is a 4NHP compound with piston valves and weighs seven tons. It uses the same round 'stayless' firebox design as the same companies tandem road rollers and wagons.

46. Ruston & Hornsby 4NHP 'Lincoln Imp' tractor number 52573 'Lucifer' was built in 1918, spending a time as works engine before going to the Isle of Man where it was used for dock haulage duties. In 1929 it was converted to a roller, being returned to tractor form in the early sixties.

47. Ruston & Hornsby of Lincoln built a large number of engines for the War Department during the First World War. Number 52607 was one of these being completed in 1918. Three years later it was sold to Courtaulds and used on coal haulage; after changing hands again it was used on agricultural work. This engine is also one of the famous 'Lincoln Imp' design, so named after the door knocker found at the great Minster in the City of Lincoln.

48. This fine Sentinel 'DG' timber tractor was in use until 1956. The engine is works number 9097 built in 1933, seen here at the 1986 Bishops Castle rally, when a number of Sentinels attended to mark the 25th. rally; many undertaking a road run from their birthplace at Shrewsbury.

49. Looking at this superbly restored Sentinel 'Super' tractor it is difficult to believe that it was once derelict, which was the case here as the engine was discovered in a scrapyard. The Sentinels works number is 6504, built at Shrewsbury in 1926. The wheelbase indicates that it most likely started life as a wagon.

50. Another earlier photograph of Sentinel 6504 taken at Haddenham in 1977, showing the other side and matching trailer. The Sentinel has travelled extensively since restoration, including on the Continent.

51. This unusual Sentinel 'Super' tractor is named 'The Elephant'. For many years this engine shunted railway wagons at Teignmouth Quay in Devon. In 1989 the engine, works number 5644 built in 1924, was sold to a Dutch enthusiast.

52. Sentinel timber tractor 'Brutus' was built in 1933, this engine is fitted with two 12NHP engines, one for propulsion, the other powering the winch. This fine Sentinel is part of the Bressingham collection situated near Diss in Norfolk.

53. Taskers of Andover, Hampshire named this neat design 'The Little Giant' class A1. Three of these three ton tractors are known to survive, this one, number 1318 was built in 1906. The A1 class is a 4NHP single cylinder design. The engine was supplied new to an owner in Kent together with a 2 ton capacity trailer; these were used for carting coal from the station to a large house. In 1930 the engine changed hands and was used on timber work in Hampshire.

54. Tasker type B2 chain driven tractor number 1895, was built in 1922. The engine was photographed at the 1989 'Great Dorset' Steam Fair'. Several Taskers of various types were present at this event.

55. This fine example of the B2 'Little Giant' type 5NHP tractor works number 1902, was built at Basingstoke in 1923. The engine is seen here at the 1990 Little Wymondley rally.

56. Another view of Tasker 1902 of 1923, this time showing the opposite side of this smart 5NHP tractor as it is made ready for a busy weekend. Note the large brass Taskers worksplate.

57. 'Goliath' is a fine example of the Wallis & Steevens 3NHP tractor, built in 1904. The early history of this tractor is unknown; it did however spend a period of its working life at a brickyard near Reading. The engine is fequently seen at the Dorset Steam Fair where this photograph was taken.

58. Many of the surviving Wallis & Steevens tractors are of the oil-bath design, including this one, number 7640 built at Basingstoke in 1919. Note the type of belly tank fitted to these tractors.

59. Wallis & Steevens 'Oil-bath' 4NHP tractor, works number 7871 was built in 1926 and is seen here at the 1962 Raynham, Norfolk rally. Note the Wallis & Steevens transfer on the belly tank.

60. This six ton tractor was originally built as an articulated unit and later rebuilt as a ballast tractor. Number 2008 was built by the Yorkshire Patent Steam Wagon Company in Leeds, in 1927.

Finale!   A 'full frontal' of 'Perseverance' showing clearly the famous Foden badge and crest.   This engine has been used in its lifetime on many duties including timber haulage in South Wales.   The engine is works number 13068 built in 1928 and is regularly seen at events in Eastern England.